FLORA OF TROPICAL EAST AFRICA

RHIZOPHORACEAE

JOHN LEWIS

Shrubs and trees. Leaves simple, usually opposite and persistent; stipules interpetiolar, caducous or absent. Flowers usually bisexual (rarely monoecious) regular, axillary, solitary to numerous in dichotomous cymes, racemes of spikes or fascicles. Calyx connate, 4–15-partite, valvate in aestivation, persistent. Petals equalling calyx-lobes in number, distinct, often clawed, sometimes fleshy or folded, frequently divided above. Stamens perigynous or epigynous, 2–4 times as many as the calyx-lobes, usually in one whorl, sometimes in pairs opposite the petals, variously inserted in relation to the disc if present; filaments sometimes very short; anthers introrse, characteristically 4-celled and dehiscing longitudinally (with numerous pollen sacs, dehiscing irregularly in *Rhizophora*). Ovary syncarpous, superior to inferior, (1–) 2–4 (–5)-celled; placentae axile, each with usually 2 pendulous anatropous ovules (1 in *Anisophyllea*); stigma usually divided. Fruit a berry (rarely dehiscent), rarely a drupe or a dry septicidally dehiscent 2–4-celled capsule. Seeds usually albuminous (not in *Anisophyllea*), sometimes arillate, often viviparous [1]; embryo straight and often with green cotyledons.

A widespread tropical and mainly Old World family including several species which dominate parts of the mangrove [2] formations of muddy shores. Of moderate economic importance: a few of the trees providing good timber, some particularly for underwater construction work; the bark of some yielding tannins, and the mangrove species in our area supplying firewood and charcoal locally. The poles of the taller mangroves are exported to the southern Middle East by Arab traders.

Plants viviparous, hypocotyls growing from fruit while on the tree; flowers solitary and pedicellate or in cymes with evident peduncles (in our area); plants of mangrove formations in intertidal areas on muddy shores and estuaries:

Flowers c. 1 cm. or less long, in cymes; calyx-lobes 4 or 5 (–6); hypocotyl thicker distally and ± sharply pointed:

Leaves elliptic; calyx-lobes 4 (–5); hypocotyl not longitudinally ridged . . . . 1. **Rhizophora**

Leaves obovate; calyx-lobes 5 (–6); hypocotyl distinctly longitudinally ridged . . . 2. **Ceriops**

Flowers more than 1·5 cm. long, solitary; calyx-lobes 15; hypocotyl ± uniform in diameter or thicker medially and blunt . . . . . 3. **Bruguiera**

[1] Simple accounts of the mechanism of vivipary are given by A. B. Rendle, Class. Flw. Plts. 2 : 378 (1925) and by Salvoza in Natural appl. Sci. Bull. 5 : 180 (1936).
[2] The ecology of mangroves (in the New World) is well covered by J. H. Davis in Papers from Tortugas Lab. 32 : 307 (1940), where an extensive bibliography can be found. Information on the cultivation and economic uses of mangrove trees can be found in Brown and Fisher's " Philippine Mangrove Swamps," Bull. No. 17 of Philipp. Dept. of Agriculture (1918). A significant account by W. Troll of the functional anatomy of the aerial roots of mangroves (incl. *Bruguiera*) is in Ber. dtsch. bot. Ges. 1930, Gen. Versamml. : 81 (1930).

Plants not viviparous, germination orthodox ; flowers
    on spikes or (very) shortly pedicelled in fascicles ;
    plants not of mangrove formations, not growing
    near the sea (in our area) :
    Leaves opposite, venation pinnate ; flowers in
        fascicles ; fruit capsular   .   .   .   .   **4. Cassipourea**
    Leaves alternate with 2–4 strong lateral veins
        paralleling the midrib ; flowers on spikes ;
        fruit drupaceous   .   .   .   .   .   **5. Anisophyllea**

1. RHIZOPHORA

L., Sp. Pl., 443 (1753) & Gen. Pl., ed. 5 : 202 (1754) ; Salvoza in Natural
appl. Sci. Bull. 5 : 179 (1936)

Trees or shrubs of muddy shores and estuaries, with stout opposite branches.
Aerial roots present as prop-roots and (adventitiously) developed from upper
nodes. Leaves evergreen, opposite, petiolate, entire, leathery and glabrous.
Inflorescences cymose ; bracteoles paired and persistent. Calyx connate
below, adnate to the ovary, 4 (–5)-partite, persistent, becoming reflexed.
Stamens 8–12 in a single whorl ; anthers narrowly oblong, longer than the
filaments, basifixed, triquetrous, having numerous pollen sacs within a
membranous epidermis. Ovary inferior to half inferior, 2-celled ; placentae
2-ovulate. Fruit a leathery indehiscent berry. Seed usually solitary (very
rarely 2), viviparous ; endosperm absent. Hypocotyl terete, swollen
subdistally, pointed. Embryo falling by separation from the cotyledons,
leaving the fruit on the tree.

Genus widespread throughout the world on muddy tropical shores.

R. mucronata *Lam.*, Encycl. Méth. Pl. de Bot. t. 396 (1797) & Encycl.
Méth. 6 : 189 (1804) ; DC., Prodr. 3 : 32 (1828) ; F.T.A. 2 : 407 (1871) ;
Salvoza l.c. 213 (1936) ; T.T.C.L. 472 (1949). Type : Encycl. Méth. Pl.
de Bot., t. 396

Glabrous tree up to 25 m. (known in our area up to 12 m.) tall ; bark
reddish-brown. Leaves dark green, cork-dotted beneath ; blade elliptic,
8 × 4 to 18 × 9 cm., apex mucronate, mucro up to 5 mm. long ; petiole up
to 3 (–4) cm. long. Inflorescences up to 8-flowered in our area (12, rarely
24 elsewhere) ; peduncles up to 4 cm. long ; bracteoles orbicular, apex
truncate, 2 mm. long ; buds ellipsoid-ovoid, 12 × 5 mm., subtetragonous.
Calyx cream, glabrous ; tube 2 mm. long ; lobes 4 (rarely 5), ovate, ultimate
apex obtuse. Petals cream or yellowish-white, c. 8 mm. long, fleshy, oblong,
involute, villous marginally, apex acute. Stamens 8 ; filaments 1 mm. long ;
anthers c. 6 mm. long, shortly and acutely apically appendaged. Ovary
ovoid, conical above ; style c. 2 mm. long ; stigma bifid. Hypocotyl up to
20–40 (–90) cm. long while on tree. Fig. 1.

KENYA. Kilifi creek, 15 Mar. 1945, *Jeffery* K136 !
TANGANYIKA. Tanga District : Kigombe beach, 11 July 1953, *Drummond & Hemsley*
    3245 !
ZANZIBAR. Unguja Ukoo, 5 Feb. 1929, *Greenway* 1354 !
DISTR. K7 ; T3, 6, 8 ; Z ; P ; shores of eastern Africa from near Massawa in the Red
    Sea to Durban in South Africa ; Madagascar ; Seychelles ; widely spread on Old
    World tropical shores from the east African coast to the islands of Polynesia (north
    in the Indian Ocean to the mouth of the Indus and in the Pacific Ocean to Okinawa
    Island, East China Sea ; south to Stradbroke Island, Queensland, Australia)
HAB. Intertidal mud-flats of shores and estuaries ; common on the seaward side of
    mangrove formations ; sea level
VARIATION. The infraspecific variants recognized by Salvoza are scarcely worthy of
    taxonomic distinction. Whatever status they may be given in the future, the material
    from our area will remain in the taxon " *mucronata*."
NOTE. The name *R. candelaria* has been misapplied to our species in the eastern part
    of its range and by Richard, Voy. Abyss. 4 : 271 (1847). The synonymy may be
    consulted in Salvoza's monograph.

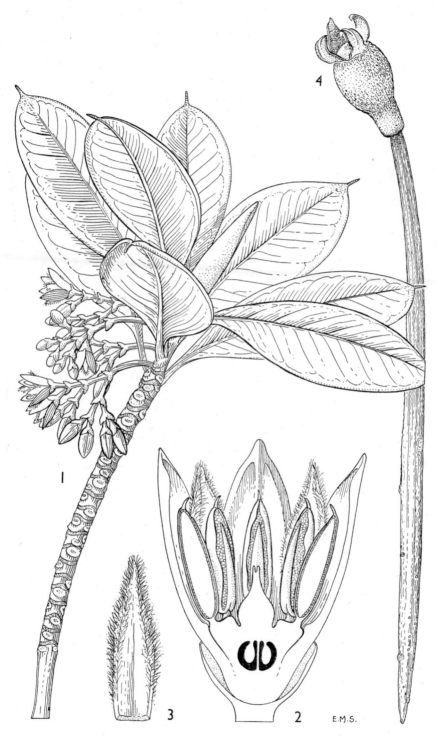

FIG. 1. *RHIZOPHORA MUCRONATA*—**1,** flowering branch with lower leaves removed, × ⅔ ; **2,** L.S. flower, × 2 ; **3,** petal, × 4 ; **4,** fruit and hypocotyl, × ⅔. 1 from *Schlieben* 2628 ; 2, 3 from *Elliott* 263; 4 from *Greenway* 4861.

FIG. 2. *CERIOPS TAGAL*—**1,** flowering branch with some leaves removed, × 1 ; **2,** underside of leaf, × 2 ; **3,** fused bracteoles, × 3 ; **4,** bud with bracteoles, × 3 ; **5,** flower, × 3 ; **6,** petal, × 6 ; **7,** pistil and stamens, × 6 ; **8,** stamen, × 6 ; **9,** fruit and hypocotyl (inverted), × ½.   9 from *Toms* 65.

2. CERIOPS

Arn. in Ann. Nat. Hist. 1 : 363 (1838)

Small or medium-sized trees or shrubs of muddy sea-shores and estuaries ; bark grey-brown. Roots and lower part of trunk forming a ± pyramidal base. Aerial roots present as upright projections from the mud. Leaves evergreen, opposite, petiolate, entire, leathery and glabrous. Inflorescences cymose, shortly pedunculate ; bracteoles paired and persistent. Calyx connate below ; tube adnate to the ovary ; lobes 5 (–6), persistent and spreading in fruit. Petals 5 (–6), membranous, appendaged. Stamens 10 (–12), inserted on the epigynous disc ; anthers dehiscing longitudinally, introrse. Ovary inferior, 3-loculate ; style thickened below ; stigma minutely 3-fid. Fruit a leathery berry, all but one seed usually aborting. Seed viviparous. Embryo falling by separation from the cotyledons, leaving the fruit on the tree.

Genus common on the tropical shores of the Indian Ocean and thence eastwards. It comprises only one species, which has been known by many names (e.g. north of our area as *C. somalensis*, and south as *C. mossambicensis*) ; the full synonymy will be given in a forthcoming paper in K.B.

C. tagal (*Perr.*) *C. B. Robinson* in Philipp. Jour. Sci. 3 : 306 (1908) ; T.T.C.L. 472 (1949). Type not seen ; probably from the southern Philippine Islands.

Glabrous shrub or small bushy-topped tree up to 6 m. tall ; bark grey-brown. Leaves yellow-green ; blade broadly elliptic to obovate, 3–9·5 cm. long, 1·4–4·8 cm. broad, apex rounded, base cuneate ; petiole up to 2·5 cm. long (in Africa). Inflorescence 4–8-flowered ; peduncle ± 1 cm. long ; bracteoles deltoid, apex obtuse, 2 mm. long ; buds ovoid, pentagonous. Calyx-tube ± 1·5 mm. long, reddish brown ; lobes 5, pale green, triangular, ± 4 mm. long, extreme apex obtuse or acute. Petals 5, white, oblong, ± 3 × 1 mm., glabrous, truncate apically and bearing (in our area) 3 distinctly clavate appendages. Disc composed of 5 ± separate lobed elements. Stamens 10 ; filaments white, 3·0 mm. long ; anthers ochre, 0·3 mm. long, basifixed. Ovary pale green ; placentae axile, biovulate. Hypocotyl extended up to 25 cm. long from fruit, sharply longitudinally ridged, ± swollen subdistally, sharply pointed. Fig. 2.

KENYA. Mombasa, Port Tudor, Dec. 1931, *MacNaughton* 129 *in Forestry Dept.* 2715 !
TANGANYIKA. Pangani District : Bushiri Estate, 24 Apr. (fl.), and 15 May (fr.) 1950, *Faulkner* 570 ! ; Uzaramo District : Dar es Salaam, 26 Feb. 1926, *B. D. Burtt* 202 !
ZANZIBAR. Zanzibar Island, Mbwani, 4 Feb. 1929, *Greenway* 1340 !
DISTR. **K**7 ; **T**3, 6 ; **Z** ; **P** ; Somaliland ; Portuguese East Africa ; Madagascar and widely ranging in the East (as far as New Ireland, Bismarck Arch., New Guinea) like *Rhizophora mucronata*, but strictly tropical in the northern hemisphere and only slightly subtropical (on the Australian west coast) in the southern.
HAB. Intertidal mud-flats and estuaries. Locally common on shoreward side of mangrove formations below high-water mark, but extending up to a mile inland in creeks ; not found elsewhere than saline creeks and mud-flats ; sea level

SYN. *Rhizophora tagal* Perr. in Mém. Soc. Linn., Paris 3 : 138 (1825)
　　C. candolleana Arn. in Ann. Nat. Hist. 1 : 364 (1838) ; F.T.A. 2 : 409 (1871) as *C. candolliana*. Type : NW. Australia, Careening Bay, *Cunningham* (K, lecto. !)
　　C. boviniana Tul. in Ann. Sci. Nat., sér. 4, 6 : 112 (1856) ; Arènes, Fl. Madag., Rhizophor. : 36 (1954). Type : Madagascar, Suareziani Bay, *Boivin* 2689 and other specimens (? P, syn.)

VARIATION. C. T. White's var. *australis*, described in J.B. 64 : 220 (1926) is a minor probably developmental variant, which has a very short unridged hypocotyl and is found only at the limit of the species' range in Western Australia. Petals with fimbriate apices are known from the eastern part of the range and may occur in our area.

3. **BRUGUIERA**

Lam., Encycl. Méth. 4 : 696 (1796)

Medium-sized trees of muddy sea-shores and estuaries. Roots and lower part of trunk forming a pyramidal buttressed base. Aerial roots present as elbowed arches arising from the mud. Leaves evergreen, opposite, petiolate, entire, leathery and glabrous. Flowers solitary or few, pedunculate, ebracteolate. Calyx connate below, adnate to the ovary, 8–15-partite, glabrous, persistent, rarely becoming reflexed (not in our species). Petals as many as the calyx-lobes, inserted at the mouth of the calyx-tube, bifid, lobes setaceous. Stamens twice as many as the petals, enfolded in pairs within them, dehiscing longitudinally, introrse. Ovary inferior, 2–4-celled ; ovules axile in pairs ; stigmatic lobes minute. Fruit a leathery berry bearing the persistent calyx-lobes ; all but one seed aborting ; pericarp accrescent. Seed viviparous. Embryo falling from the tree with the fruit attached.

Genus widespread on Old World tropical shores.

B. gymnorrhiza[3] (*L.*) *Lam.*, Encycl. Méth., Bot. 4 : 696 (1796) ; Encycl. Méth. Pl. de Bot. t. 397 (1797) ; T.T.C.L. 471 (1949). Type : Rheede, Hort. Malab. 6 : t. 31 (lecto !)

Glabrous tree up to 12 m. tall (nearer half this height in our area), bark reddish-brown. Leaves dark green ; blade narrowly to broadly elliptic, 5–15 cm. long, subacuminate above to an acute apex, cuneate below to an acute base, petiole up to 3 cm. long. Flowers solitary ; peduncles recurved, about 1 cm. long ; buds narrowly ellipsoid, terete, c. 25 × 10 mm. Calyx pinkish-green to reddish-brown ; tube 1–2 cm. long ; lobes (10–) 12 (–15), linear, 1·8–2·0 cm. long, apex acute. Petals white, soon turning brown, about 1·5 cm. long, conduplicate below, bifid with a median seta, stiffly pubescent, especially below ; lobes 3–3·5 mm. long, (bi- or) tri-ciliate apically. Anthers 4 mm. long, dorsifixed, mucronate above ; filaments 6 or 8 mm. long (one shorter and one longer in each pair). Ovary 3-celled. Fruit turbinate, 2 cm. long. Hypocotyl up to 11 cm. long on tree, terete or only shallowly and bluntly longitudinally ridged, swollen ± uniformly throughout length. Fig. 3.

KENYA. Mombasa, 19 Aug. 1949, *Bogdan* 2633 !
TANGANYIKA. Rufiji District : Mafia Island, Ras Mbisi, 2 July 1932, *Schlieben* 2614 !
ZANZIBAR. Unguja Ukoo, 5 Feb. 1929, *Greenway* 1353 !
DISTR. **K**5 ; **T**3, 6, 8[4] ; **Z**, **P** ; Portuguese East Africa, Madagascar, South Africa and widely ranging to the East like *Rhizophora mucronata* but neither extending so far beyond the tropics nor so far into the Pacific Ocean (but see Note below)
HAB. Intertidal mud-flats and estuaries ; on the less exposed parts ; sea level

SYN. *Rhizophora gymnorrhiza* L., Sp. Pl. 443 (1753) ; DC., Prodr. 3 : 33 (1828)
     R. conjugata L., Sp. Pl. : 443 (1753) but not of other authors. Type : P. Hermann Herb., Illustr. No. 279 (BM, holo. !)
     Bruguiera cylindrica Blume as used by Oliver in F.T.A. 2 : 409 (1871) at least in part ; see Note.
     B. capensis Blume in Mus. Bot. Lugd. Bat. 1 : 137 (1827). Type : South Africa, *Drege* (K, ? iso. !)

NOTE. *R. cylindrica* L. is the plant commonly called *B. caryophylloïdes* in the East Indies. The above conception of *B. gymnorrhiza* excludes *B. sexangula* (Lour.) Poir. (type from Indo-China, *Loureiro*—BM, iso. !), which differs only in having sometimes fewer calyx-lobes, yellow flowers and slightly narrower leaves. Blume's conception of *B. cylindrica* appears to have included this Eastern taxon. *R. conjugata* is reported as having red flowers in the East. A future revision may well include *B. sexangula* as an infraspecific variant within our species, which conception would give *B. gymnorrhiza* a much wider distribution in the Pacific Ocean.

[3] Linnaeus' epithet with a single "r" is an orthographic error, which was first corrected by Lamarck.
[4] No specimens seen from Southern Province. Record kindly provided by Dr. P. J. Greenway.

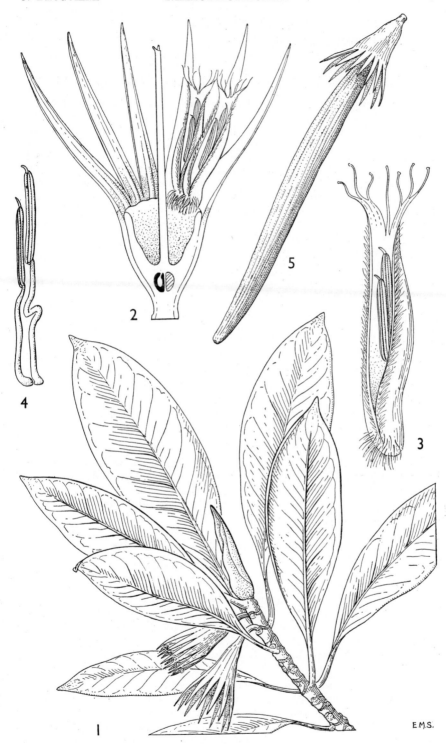

FIG. 3. *BRUGUIERA GYMNORRHIZA*—**1,** flowering branch with some leaves removed, × ⅔ ; **2,** L.S. flower with all but two petals and stamen-pairs removed, × 3 ; **3,** petal with enclosed stamen-pair, × 6 ; **4,** stamen-pair, × 6 ; **5,** fruit and hypocotyl, × ⅔.   1, 5 from *Greenway* 5287 ; 2–4 from *Johnson* 22.

4. CASSIPOUREA

Aubl., Pl. Guian. 1 : 528 (1775) ; Alston in K.B. 1925 : 241 (1925)
Richaeia Thou., Gen. Nov. Madag. 25 (1808) [5]
Weihea Spreng., Syst. 2 : 594 (1825) [5]
Dactylopetalum Benth. in J.L.S. 3 : 79 (1858)

Small shrubs to large trees, without aerial roots. Leaves decussate or rarely verticillate, elliptic to more rarely obovate or ovate, rarely circular, leathery to membranous, very usually petiolate, entire to serrate, especially towards the apex. Flowers solitary to numerous and fascicled ; pedicels articulate. Calyx connate, sometimes adnate to the ovary, variously campanulate, 4–7-lobed ; lobes valvate. Petals inflexed in bud, margin divided above (usually laciniate), usually glabrous. Stamens 8–45, variously inserted in relation to the disc (which may be very little developed or ? absent). Ovary superior, half-inferior or sub-inferior, 2–4-celled ; style often persistent. Fruit a thinly fleshy septicidally dehiscent capsule ; seeds 2–4, arillate, with a leathery testa ; endosperm fleshy ; embryo straight with flat cotyledons.

A tropical and south subtropical genus mainly of inland forests, on river banks, and (*C. barteri* in W. Africa) by sandy sea-shores. It includes a few useful timber trees. Fairly generally distributed from British Guiana in the west to Ceylon in the east.

Cassipourea in the sense here adopted includes the three genera *Weihea*, *Dactylopetalum* and *Lasiosepalum* which were reduced to subgenera by Alston in his monograph, K.B. 1925 : 241–276 (1925). Members of two of these subgenera occur in our area and they are diagnosed below ; of the other two subgenera one occurs in West Africa and the other in tropical America. Changes in the specific limits have been made by the present author for this Flora only and the taxonomic treatment of the genus as a whole is in a state of flux ; see K.B. 1955 : 143 (1955).

KEY TO THE SUBGENERA

Leaves usually ± hairy ; calyx divided to beyond
    the middle, ± densely hairy externally   .   subgenus *Weihea*
Leaves glabrous ; calyx not divided to the
    middle, glabrous or nearly so externally   .   subgenus *Dactylopetalum*

Cassipourea subgenus **Weihea** (*Spreng.*) *Alston*

in K.B. 1925 : 250 (1925)

Leaves very usually hairy, sometimes obscurely so. Flowers bracteolate. Calyx very shortly campanulate ; lobes spreading, longer than the tube (twice as long or more), hairy externally, glabrous or puberulous within. Stamens three or more times as many as the calyx-lobes. Ovary 3 (–4)-locular.

A subgenus of tropical and southern Africa, Mascarene islands, and with one species in Ceylon.

Leaves puberulous on both surfaces (obscurely so
    above) ; calyx more than 1 cm. long, distinctly
    and ± densely puberulous within ; stamens
    more than 25   .   .   .   .   .   .   1. *C. mollis*

[5] *Weihea* Spreng. is conserved against *Richaeia* Thou. in Int. Code Bot. Nomen. 122 (1952).

Leaves usually glabrous above and other than puberulous beneath ; calyx less than 1 cm. long, glabrous or sparsely hairy within ; stamens less than
25 :
    Leaves (all) rotund or subcircular, usually 2·5 to
    4 cm. long and margin usually entire . . 2. *C. rotundifolia*
    Leaves not rotund or subcircular (a few may
    approach these shapes), mostly more than 4 cm.
    long or, if less, then margin divided, at least
    above :
      All leaves (in our area) 10 cm. long or less :
        Plants with nearly all leaves obovate, clearly
        broadest above the middle . . . 3. *C. celastroïdes*
        Plants with all leaves narrowly to broadly
        elliptic, not broadest above the middle :
          Leaf-margins all entire or mostly entire with
          a few widely serrate-dentate ; ovary
          densely and shortly hirsute . . 4. *C. euryoïdes*
          Leaf-margins very usually (shallowly)
          serrate, especially distally ; ovary
          glabrous or sparsely and longly hirsute,
          especially above . . . . 5. *C. malosana*
      All or some leaves more than 10 cm. long :
        Ovary glabrous or very sparsely yellow-hirsute
        above ; fruit ± 3-angled, truncate above,
        smooth, glabrous and black . . . 6. *C. congoensis*
        Ovary densely white-hirsute ; fruit not angled,
        rounded above, verrucose, densely pubes-
        cent and yellow brown . . . 7. *C. ruwensorensis*

1. **C. mollis** (*R. E. Fries*) *Alston* in K.B. 1925: 257 (1925). Type: Northern
Rhodesia, Abercorn, *R. E. Fries* 1251 (LD, syn.) & Kalambo, *R. E. Fries*
1251a (B, syn. †, BM & K, photo-syn. !)

Shrub up to 5 m. tall ; young stem ± densely puberulous, only weakly
glabrescent ; one year old stem distinctly puberulous. Leaves petiolate ;
petiole up to 1 cm. long, hirsute ; blade elliptic, ovate to broadly ovate
or subrotund to broadly obovate, up to 10 × 8 cm., rounded or cuneate
below to an acute base, rounded to subcuneate above, apex truncate to
rectangular, margin entire or shallowly and sharply serrate, midrib and
nerves hirsute below, otherwise puberulous at maturity on both surfaces
(sometimes shortly and obscurely so above). Flowers precocious, usually
solitary, smelling of cucumber salad ; pedicels up to 6 mm. long, ± spreading-
hirsute, articulate above. Calyx (4–) 5-partite ; tube 2–5 mm. long ; lobes
7–10 (–15) mm. long, ± appressed-pubescent externally, ± densely appressed-
white-puberulous within. Petals (4–) 5. Stamens up to (26–) 35–40 (–45),
very closely set on the margin of a 2 mm. long disc. Ovary superior, 3-celled,
very densely and regularly pilosulous ; style up to 5 mm. long. Capsule
ovoid, 12–15 mm. long, densely puberulous.

TANGANYIKA. Tabora, Simbo Reserve, June 1938, *Wigg* 1098 ! ; Dodoma District :
near Manyoni, between Mkwesi and Kunguya, 18 Dec. 1931, *B. D. Burtt* 3450 !; [N.]
Kilosa District, Kibedya, Jan. 1931, *Haarer* 1981 !
DISTR. T1, 4–6 ; Northern Rhodesia
HAB. A more important constituent of *Brachystegia* (-*Isoberlinia*) woodland and in
" Commiphora bush " on rocky or acid (sandy) soil, 750–1600 m.

SYN. *Weihea mollis* R. E. Fries, Schwed. Rhod. Kongo. Ex. Bot. 1 : 165 (1914)
    W. insignis Engl. in E.J. 54 : 359 (1917)

C. insignis (Engl.) Alston in K.B. 1925 : 257 (1925) ; T.T.C.L. 471 (1949).
Type : Tabora, *von Trotha* 44 (B, lecto. †, BM & K, photo-lecto. !)

Note. It is considered inadvisable to retain *C. insignis* as a distinct species even
though no authentic specimen of it has been examined. It appears from description
to be a slightly smaller variant and is said (? in error) to have only about 20 stamens.

2. **C. rotundifolia** *(Engl.) Alston* in K.B. 1925 : 256 (1925). Type :
Tanganyika, Morogoro District: Uluguru Mts., Lukwangule Plateau
Stuhlmann 9204 (B, holo. †, BM & K, photo. !)

Tree to 9 m. tall ; young stem sparsely pubescent. Leaves petiolate ;
petiole 3–4 (–5) mm. long, appressed-pubescent ; blade subcircular to
elliptic, rarely narrowly elliptic, up to 4·6 (–7·5) × 3·2 (–4) cm., broadly
tapering or rounded below to a subacute base, rounded above (rarely subcuneate) to an obtuse apex, margin usually entire (see Note), midrib longly
appressed-pubescent beneath, otherwise glabrous, somewhat leathery.
Flowers 1–3 per inflorescence ; pedicels up to 5 (–7) mm. long, spreading-
pilose, articulate towards the apex. Calyx 5-partite ; tube 2 mm. long ;
lobes 5 mm. long, very narrowly ovate (almost lanceolate), apex acute, ±
sparsely pubescent externally, glabrous within or merely minutely puberulous
on the margin only and/or at the apex. Petals 5, usually greenish-white
(see Note). Stamens c. 20, inserted on the margin of an inconspicuous disc.
Ovary half-inferior, 3–4-celled, longly and densely white-pilose ; style 2 mm.
long, persistent ; stigmas flattened. Capsule not seen.

Kenya. Nairobi, Kirichwa Ndogo valley, Mar. 1940, *Bally* 804 ! ; Nairobi Arboretum,
28 Apr. 1919, *Battiscombe* 930 !
Tanganyika. Morogoro District : Uluguru Mts., Lukwangule Plateau, *Schlieben*
3552 ! ; Njombe District : Msima Farm, 1932, *Emson* 324 !
Distr. **K**4, 6 ; **T**6, 7 ; not known elsewhere
Hab. In open or closed woodland and as a shrub relic in grassland, c. 1600–2300 m.

Syn. *W. rotundifolia* Engl. in E.J. 28 : 440 (1900)

Variation. The specimen *Emson* 324 (called *Muzaizi* in the native vernacular as is
C. malosana) although cited here, is anomalous in having the leaf margin sparsely,
widely and shallowly serrate above the middle, and is a shrub. It is certainly a close
ally of the members of this species, but may be a hybrid with *C. malosana*. The flower
is reported by Emson as pale yellow ; a recent gathering of *C. malosana* from Nairobi
Arboretum, *Williams* 390, had green-white flowers.

3. **C. celastroïdes** *Alston* in K.B. 1925 : 256 (1925). Type : Kenya,
Teita District, Ndara Mt., *Hildebrandt* 2475 (K, holo. !)

Shrub or small tree ; young stem glabrous or ± spreading-pubescent,
glabrescent ; first-year stem glabrous or shortly pubescent. Leaves petiolate ; petiole 2–4 mm. long, indumentum as on stem ; blade ± ovate or
rarely narrowly obovate, 2·5–4 × 1·5–3 cm., rounded to a truncate base or
cuneate below to an acute base, rounded or rarely subacuminate to a
truncate apex, margin ± shallowly serrate especially towards the apex or
very rarely entire, glabrous above, glabrous or sparsely pubescent beneath,
glabrescent. Flowers 1 (–3) per inflorescence ; pedicels up to 3·5 mm. long,
densely appressed-puberulous, articulation median. Calyx (4–) 5 (–6)-
partite ; tube 1·5 mm. long ; lobes 4 mm. long, narrowly oblong-triangular,
apex acute, densely appressed-pubescent externally, glabrous or very
nearly so within. Petals 5, laciniate above. Stamens c. 20 (19–22), arising
from margin of a very short disc, anthers introrse. Ovary shortly and ±
densely pilose and interspersed, especially towards the apex, with longer hairs,
becoming more sparsely hirsute ; style up to 2·5 mm. long, persistent ;
stigmas spreading. Capsule (very immature) very broadly ovoid, 3 mm.
long, ± regularly densely hirsute.

KENYA. Near Nyeri, 6 Mar. 1922 *Fries* 2103 ; Machakos District : Garabani Hill, 6 Mar. 1940, *van Someren* 15 ; Kitui District : Yatta Plateau, Moboloni Rock, 7 Dec. 1952, *Bally* 8375 !
DISTR. **K**4, 7 ; not known elsewhere
HAB. On rocky hillsides ; 600–1500 m.

VARIATION. The specimens of this species show some interesting variation. The Fries specimen has all the leaves cuneate below to an acute base and the very immature fruits rather sparsely hairy ; the holotype and *van Someren* 15 have their leaves mostly rounded below to a truncate base and their ovaries more densely hairy ; *Bally* 8375 has the leaf bases of the latter pair and the ovary indumentum of the former. *Van Someren* 15 and *Bally* 8375 are the only ones with mature petioles showing hairs.

NOTE. This species is very close morphologically to *C. flanaganii* (Schinz) Alston and only slightly less so to *C. gerrardii* (Schinz) Alston. These two species have sub-spherical pubescent fruits, and occur in South Africa ; if the mature fruits of *C. celastroïdes* are closely similar, some geographically intermediate material would throw suspicion upon the distinctness of our species.

4. C. euryoïdes *Alston* in K.B. 1925 : 254 (1925). Type : Kenya, North of Mombasa to Lamu & Witu, " Masusi," *Whyte* 22 (BM, holo. !)

Tree or shrub up to 18 m. tall ; young stem very shortly and rather sparsely appressed-hirsute, glabrate within one year. Leaves petiolate ; petiole up to 6 mm. long, sparsely hirtellous ; blade elliptic to almost narrowly (oblong-) elliptic, up to 7·8 × 3·2 cm. or 6·7 × 3·5 cm., cuneate or subacuminate to an acute base, acuminate or shortly acuminate to an obtuse apex, margin entire to widely serrate, especially above the middle, glabrous or very sparsely pubescent below, smooth and shining above, papery to chartaceous. Flowers 1–3 ; pedicels up to 1·5 mm. long, shortly hirsute, articulate apically. Calyx 5–6-partite ; tube 1·5 mm. ; lobes 3 mm. long, narrowly triangular, ± shortly white-hirsute externally, ± obscurely puberulous within. Stamens 20–24 ; disc c. 0·5 mm. long. Ovary densely white-hirsute ; style persistent, usually sparsely pubescent (? glabrescent). Fruit almost spherical, c. 5 mm. long, densely velvety ; indumentum interspersed with ± numerous longer hairs above.

KENYA. Kilifi District : Mida, 1929, *R. M. Graham* 1627 ! ; Kwale District : Shimba Hills, Mwale-Mdogo Forest, 23 Aug. 1953, *Drummond & Hemsley* 3964 ! and Buda (-Mafisini) Forest, Aug. 1936, *Dale* 3549 !
TANGANYIKA. Uzaramo District : Kerekese Forest Reserve, Sept. 1953, *Paulo* 145 !
DISTR. **K**7 ; **T**6 ; Portuguese East Africa
HAB. Coastal forest and scrub, rarely in secondary bushland ; 80–400 m.

SYN. *C. honeyi* Alston in K.B. 1925 : 254 (1925). Type : Portuguese East Africa, Siluvu Hills, *Honey* 757 (K, holo. !)

VARIATION. The East African material as a whole shows only a slight tendency to have the leaves serrate above the middle, most gatherings having entire leaves. The greatest degree of division is shown by some leaves on *Moggridge* 399 from Kilifi. The Portuguese East African specimen on the other hand has narrower leaves, mostly widely and shallowly dentate or serrate, and may represent a variety.

5. C. malosana (*Baker*) *Alston* in K.B. 1925 : 258 (1925) ; J. Lewis in K.B. 1955 : 144 (1955). Type : Nyasaland, Mt. Malosa, *Whyte* (K, holo. !)

Evergreen tree (6–) 12–45 m. tall ; young stem ± appressed-pubescent, glabrescent (first year stem scattered-hirsute). Leaves petiolate ; petiole 3–8 mm. long, indumentum similar to stem ; blade narrowly to almost broadly elliptic or more rarely obovate, usually c. 5 × 3 cm., exceptionally c. 10 × 5 cm., cuneate or subrounded to an acute base, shortly to longly acuminate or rarely rounded, extreme apex very usually obtuse, margin ± widely bluntly or sharply serrate, especially above, to rarely entire, glabrous above, ± sparsely and obscurely appressed-pubescent beneath especially on the midrib, hairs thin, lying regularly in the direction of the apex, glabrescent, papery to chartaceous. Flowers 1–5 (–8) per inflorescence ; pedicels up to 4 mm., rarely 7 mm. long, articulate above the middle, densely puberulous. Calyx 4–5 (–6)-partite ; tube 1 mm. long ; lobes 3·5–5 mm. long, narrowly oblong-triangular, apex acute, densely appressed-pubescent externally, glabrous or very nearly so within. Petals 4–5, laciniate apically. Stamens (15–) 16–20 (–22), arising from the margin of a very short disc ; anthers introrse. Ovary superior, 3 (–4)-celled, glabrous or almost so with a few apical hairs to longly hirsute especially towards the apex ; style persistent. Capsule ovoid, 6·5–8 mm. long, shortly appressed-pubescent especially above the middle, with longer upright hairs apically, glabrescent, rarely glabrate. Capsule longer than the persistent style.

UGANDA. Karamoja District : Mt. Debasien, *Eggeling* 2727! ; Acholi District : SE. Imatong Mts., headwaters of Aringa R., 7 Apr. 1945, *Greenway & Hummel* 7302!
KENYA. Baringo District : Katimok forest, Oct. 1930, *Dale* 2452!; Kiambu District : Chania Forest, 28 Nov. 1942, *Logie in Bally* 7968! ; Masai District : Ngong Forest, *Bally* 6529! & *Moon* 21!
TANGANYIKA. Masai District : SE. slope of Ngorongoro crater, 22 Feb. 1933, *Greenway* 3369! ; Arusha District : Mt. Meru, July 1951, *Parry* 65! ; Iringa District : Ihangana Forest Reserve, 23 Aug. 1937, *Pitt[-Schenkel]* 561!
DISTR. **U**1 ; **K**1, 3–6 ; **T**2, 3, 7 ; Somaliland Protectorate and Ethiopia
HAB. Upland dry forest and under-canopy of moist forest ; (1100–) 1700–2600 m.

SYN. *W. malosana* Baker in K.B. 1897 : 267 (1897)
    W. elliottii Engl. in E.J. 40 : 52 (1907). Type : Kenya, Nairobi District : Mosigi, *Elliott* 145 (K, iso. !)
    W. eickii Engl. in E.J. 40 : 50 (1907). Types : Tanganyika, W. Usambaras, plateau near Kwai, *Albers* 198, 19 & *Eick* 137 (B, syn. † ; BM & K, photo-syn. !) ; *Drummond & Hemsley* 1353 (K, EA, neo. !)
    W. ilicifolia v. Brehm. in E.J. 54 : 362 (1917). Type : Tanganyika, Mfimbwa Mt., *Fromm* 230 & 232 (B, syn. †)
    C. eickii (Engl.) Alston in K.B. 1925 : 259 (1925) ; T.T.C.L. 472 (1949)
    C. elliottii (Engl.) Alston in K.B. 1925 : 260, & fig. 243 (1925) ; I.T.U., 2nd ed., 330 (1952) ; T.T.C.L. 472 (1949)
    [*W. wambugensis* Engl. in V.E. 3 (2) : 669 (1921). Type : from Tanganyika, W. Usambaras, Wambuguland (B, †) nom. nud.]

VARIATION. The North East African material of this species, previously segregated as *C. abyssinica*, *C. avettae* and *C. salvago-raggei* [see K.B. 1955 : 144 (1955)], shows a tendency to include specimens with narrower more acuminate leaves ; material from Uganda is similar. Kenya material is usually small-leaved (terminal sprigs may bear leaves up to only 4–5 mm. long), the leaves having scarcely any acumen. These differences are not absolute however and the variation in the degree of serration of the leaf margin is not correlated with them. More material would be welcomed from south of latitude 5° S.

NOTE. *C. malosana* is an important timber tree in East Africa. It has very hard wood of good colour and is referred to in the native vernacular as " *Muzaizi*."

6. C. congoensis DC., Prodr. 3 : 34 (1828) ; Alston in K.B. 1925 : 259 (1925). Type : on Congo R., *Chr. Smith* (? P, holo.)

Shrub or small tree up to 13 m. tall ; young stem sparsely appressed-pubescent. Leaves petiolate ; petiole 3–8 mm. long, indumentum as on stem; blade narrowly to broadly elliptic, more rarely subovate or subobovate, rarely oblong, up to 14 × 5·5 cm., cuneate or rounded to a cuneate base, tapering or subacuminate to rounded above, extreme apex acute or obtuse, margin entire or sharply to bluntly serrate especially above the middle, glabrous above, blade obscurely and midrib longly appressed-pubescent or blade occasionally and midrib rarely glabrous beneath. Flowers 1–4 (–6) per inflorescence ; pedicels up to 4 mm. long, articulate just above the middle, appressed-pubescent. Calyx 5- or 4-partite ; tube 1·5 mm. long ; lobes 3 mm. long, sublanceolate, apex acute, longly and densely appressed-pubescent externally, glabrous or sparsely pubescent within. Petals 5 or 4,

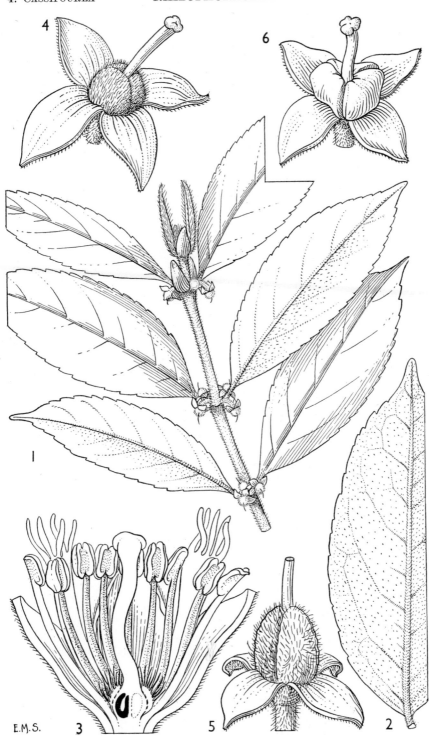

FIG. 4. *CASSIPOUREA RUWENSORENSIS*—**1,** branch, × ⅔ ;   **2,** underside of leaf, × ⅔ ;   **3,** L.S. young flower, × 8 ; **4,** young fruit, × 4 ; **5,** fruit (almost mature, ? stigma broken), × 4. *C. CONGOENSIS* —**6,** mature fruit, × 4. 1, 2, 4 from *Drummond & Hemsley* 4571 ; 3 from *Eggeling* 3171 ; 5 from *Dale* 3131 ; 6 from *Greenway & Eggeling* 7093.

shortly laciniate, glabrous. Stamens 20 or 16 (–17), inserted on outer margin of a very slightly thickened annular disc. Ovary superior, drying black, 3 (–4)-celled, glabrous or with long sparse usually yellow hairs especially above ; style up to 4 mm. long, persistent. Capsule (drying) black, broader above than below, 5–6 mm. long, smooth and glabrous or with a few hairs above. Fig. 4/6.

UGANDA. Kigezi District : Ishasha Gorge, 10 Feb. 1945, *Greenway & Eggeling* 7093 ! & May 1950, *Purseglove* 3435 !
TANGANYIKA. Lushoto District : W. Usambara Mts., Shume-Magamba Forest, 5 Feb. 1952, *Parry* 112 !
DISTR. U2 ; T3 ; widespread in West Africa ; Belgian Congo ; Northern Rhodesia and Nyasaland
HAB. Upland evergreen forest ; 1300–1950 m.

SYN. *C. africana* Benth. in Hook., Niger Fl. 341 (1849). Type : Nigeria, on the Niger R., " opposite Stirling [I.]," *Vogel* (K, holo. !)
    Weihea africana (Benth.) Oliv. in F.T.A. 2 : 410 (1871)

VARIATION. No specimen has been seen in which *all* the leaves are glabrous on the blade beneath. The ovaries on the Northern Rhodesian specimens, e.g. Mwinilunga District, *Milne-Redhead* 2536, show the greatest degree of density of indumentum, they are distinctly, although not densely, long yellow-hirsute, especially, but not only, above. The Tanganyika specimen has galled flowers and is only uncertainly placed here ; some sterile material from Kenya (e.g. *Bally* 7658) may belong here.

7. C. ruwensorensis (*Engl.*) *Alston* in K.B. 1925 : 263 (1925) ; J. Lewis in K.B. 1955 : 146 (1955). Type : Belgian Congo, West Ruwenzori, Butagu Valley, *Mildbraed* 2515 (B, holo. †) & Fort Beni, *Mildbraed* 2245 (B, neo. †, BM & K, photo-neo. !)

Tree to 12 m. tall ; young stem densely longly yellow-pubescent, darkening and glabrescent. Leaves petiolate ; petiole up to 1 cm. long, indumentum as on stem ; blade narrowly to broadly elliptic, narrowly obovate to sub-lanceolate or more rarely lanceolate, up to 15 × 7 cm., cuneate to an acute base, apex acuminate, margin broadly bluntly or sharply sometimes shallowly ± serrate especially above the middle, glabrous above, ± longly pubescent beneath especially on midrib. Flowers (2–) 4–8 per inflorescence ; pedicels up to 5 mm. long, articulate above the middle, longly and densely hirsute. Calyx 4 (–5)-partite ; tube 1·5 mm. long ; lobes 4 mm. long, triangular, acute at apex, longly and densely appressed yellow-hirsute externally, sparsely pubescent within. Petals 4 (–5), laciniate, glabrous. Stamens 16 (–17) –20 ; disc slight. Ovary half-inferior, 3–4-celled, densely white-shining-hirsute ; style up to 4 mm. long, persistent. Capsule ovoid, up to 6 mm. long, verrucose, shortly densely hirtellous and longly spreading-white- or yellow-hirsute. Fig. 4/1–5.

UGANDA. Bunyoro District : Budongo (NW. of Masindi), Siba Forest, 17 Apr. 1945, *Greenway & Eggeling* 7348 ! ; Ankole District, Bunyarunguru country, Kalinzu Forest, Aug. 1936, *Eggeling* 3171 ! ; Busoga District : 10 m. E. of Jinja, 20 Oct. 1950, *Kiragga* 3 !
KENYA. N. Kavirondo District : Kakamega, May 1933 (fr.) *Dale* 3131 !
TANGANYIKA. Bukoba District : Minziro Forest, July 1950, *Watkins* 465 *in F.H.* 3204 !
DISTR. U2–4 ; K5 ; T1 ; Belgian Congo and Ethiopia
HAB. Upland evergreen forest, sometimes in dense semi-swamp forest (Watkins) or in elephant-grassland ; 1200–1700 (–2500) m.

SYN. *Weihea ruwensorensis* Engl. in Z.A.E. 579 (1913) ; E.J. 54 : 364 (1917)
    W. boranensis Chiov. in Cufodontis, Miss. Biol. Borana 143 & photo. 144 (1939). Type : Ethiopia, Mega, *Cufodontis* 644 (FI, holo. !)

NOTE. This species is close to *C. congoensis*, being distinguished absolutely from that species only by the nature of the ovaries and fruits. The two species differ slightly in habitat and have almost distinct geographical ranges.
    A gathering from Kenya, " Nandi & Kakamega," *Wye* 1753 *in CM* 13636, of which the main bulk is unmistakably this species, has one twig on the Coryndon Museum

sheet which, although the calyx lobes are sparsely pubescent within and all the styles are not glabrous (" *ruwensorensis* "), has flowers with glabrous ovaries (" *congoensis* "). On the present system, hybridity is assumed to account for the nature of this twig within what is therefore a mixed gathering.

The Ethiopian plant is only a more broadly leaved form.

Cassipourea subgenus **Dactylopetalum** (*Benth.*) *Alston* in K.B. 1925 : 271 (1925) ; J. Lewis in K.B. 1955 : 147 (1955)

Leaves glabrous. Calyx elongate-campanulate ; lobes suberect, shorter than the tube (half as long or less). Stamens twice as many as the calyx-lobes. Ovary 2 (rarely 3) -locular.

Subgenus confined to tropical and south subtropical Africa and the Mascarene islands ; including only one species in our area.

8. **C. gummiflua** *Tul.* in Ann. Sci. Nat., sér. 4, 6 : 123 (1856) ; Alston in K.B. 1925 : 272 (1925) ; J. Lewis in K.B. 1955 : 147 (1955). Type : Madagascar, Nossi-bé Island, *Boivin* (P, holo. !, K, ? iso. !)

Tree or shrub up to 25 m. tall ; young stem very shortly appressed-pubescent, glabrescent (one-year-old stem glabrous). Leaves petiolate, petiole 6–15 mm. long, glabrous ; blade varying from narrowly elliptic to almost broadly oblong and elliptic-lanceolate to broadly elliptic, up to 11·5 × 6 cm. or 14 × 5 cm., rounded or cuneate to an acute base, rounded to acuminate or very rarely cuneate to an obtuse or acute apex, margin entire or shallowly sinuate or bluntly and shallowly serrate especially above, glabrous. Flowers numerous ; pedicels 1·5–4 mm. long, glabrous, articulate medially or above. Calyx 4–5 or 5–6 (rarely 7 [–8])-fid ; tube 2–3 mm. long, lobes 1 mm. long, semi-circular to deltoid, apex obtuse (to acute), glabrous. Stamens 8–10 or 10–12 (–14), inserted on the ± lobed disc. Ovary superior to partly inferior, very usually 2-celled (very rarely 3-celled), glabrous, ± sparsely pilose above the middle, to regularly pubescent or densely white-hirsute especially above the middle. Capsule ellipsoid or subspherical to obovoid, up to 7 mm. long, glabrous or shortly pubescent.

DISTR. **U**2, 4 ; **K**4 ; **T**6, 7 ; Southern Nigeria ; Brit. Cameroons ; S. Tomé ; Belgian Congo ; Angola (extreme north) ; Nyasaland ; Northern Rhodesia ; Portuguese East Africa ; Madagascar ; Seychelles (? introduced) and South Africa (Natal)

VARIATION. The characters of variational interest in the species as a whole are the ovary indumentum, the pedicel length, and the shape of the leaves, especially the last two. Collectors should try to assess and preserve the variation of these characters within local populations.

NOTE. Two varieties of this species occur in our area ; their diagnostic features are italicized in the descriptions below. As accepted by the present author, *C. gummiflua* Tul. is a very variable and wide-ranging species, the full treatment of which remains to be worked out. In the present arrangement the type variety is found only in Madagascar, and the fourth variant is var. *mannii* (Hook. f. ex Oliv.) J. Lewis from West Africa. There is an urgent need for specimens showing stages in the development of the flowers and fruits, especially for the latter.

var. **ugandensis** (*Stapf*) *J. Lewis* in K.B. 1955 : 159 (1955). Type : Uganda, Toro District, Kibale Forest, *Dawe* 499 (K, holo. !)

Leaves narrowly to broadly elliptic, rounded to cuneate, extreme base acute, entire to shallowly sinuate or bluntly and shallowly serrate above the middle. *Pedicels not more than 2 mm. long.* Calyx-lobes and petals 4–5 ; stamens 8 or 10. *Ovary glabrous or very sparsely and irregularly hirsute.* Ripe fruit not seen, probably glabrous.

UGANDA.   Toro District : Ruimi R., near Rwagimbo, Aug. 1943, *Eggeling* 5424 ! ;
   Kigezi District : Kayonzo Forest, Oct. 1940, *Eggeling* 4204 ! ; Masaka District :
   Namalala Forest, Nov. 1913, *Fyffe* 29 & 75 !
KENYA.   Kiambu District : south Aberdares, Chania Forest, Matara, 26 May 1942,
   Bally 7972 !
DISTR.   U2, 4 ; K4 ; Belgian Congo and Northern Rhodesia
HAB.   Rain- and swamp-forest ; 1200–1500 m.

SYN.   *Dactylopetalum ugandense* Stapf in J.L.S. 37 : 515 (1906)
       C. ugandensis (Stapf) Engl. in V.E. 3, 2 : 673 (1921) ; Alston in K.B. 1925 :
       272 (1925)

var. **verticillata** (*N.E. Br.*) *J. Lewis* in K.B. 1955 : 158 (1955). Types : South Africa,
Natal, near Pinetown, *Wood* 3876 (K, syn. !) and near Durban, *Wood* 4619 (K, syn. !)

Leaves narrowly elliptic, elliptic or rarely broadly elliptic, cuneate or rarely rounded
to an acute base, usually shallowly sinuate or bluntly and shallowly serrate above the
middle, rarely entire. *Pedicels more than 2 mm. (up to 4 mm.) long.* Calyx-lobes and
petals 5 or 6 ; stamens 10 or 12. *Ovary regularly pubescent in ± the upper half.* Fruit
puberulous above (shortly and perhaps obscurely).

TANGANYIKA. Iringa District : Kiwere [Kivere], 15 Mar. 1947, *Gilchrist* 97 *in F.H.*
   1926 ! Livingstone Mts., Upangwa, May 1953, *Eggeling* 6535 ! and Kigogo Forest
   Reserve, 9 Sept. 1937, *Pitt*[*-Schenkel*] 564 !
DISTR.   T6, 7 ; Nyasaland ; Portuguese East Africa ; South Africa ; Natal ; Seychelles
   (single record ; probably introduced) and Brit. Cameroons, Bamenda
HAB.   In upland forest ; 1900–2100 m.

SYN.   *C. verticillata* N.E. Br. in K.B. 1894 : 5 (1894)
       Dactylopetalum verticillatum (N.E. Br.) Schinz in Bull. Herb. Boiss. 5 : 866 (1897)
       C. redslobii Engl. in E.J. 54 : 368 (1917). Types : Tanganyika, Morogoro
          District : Nguru Mts., near Manyangu, *Bittkau in Redslob* 2981 (B, lecto. † ;
          BM & K, photo-lecto. showing dissections !) & *Redslob* 2907 (B, syn. †)
       C. verticillata f. *decussata* Engl. in E.J. 54 : 369 (1917). Type : Tanganyika,
          Ukinga, Manganyema Mt., *Goetze* 1215 (B, holo. †)

5. ANISOPHYLLEA

[R.Br. ex] Sabine in Trans. Hort. Soc. 5 : 446 (1824)

Trees, shrubs and subshrubs ; individuals bearing leaves of more than one
distinct size but not (in our area) very unlike in shape. Leaves of the
larger sort alternate, shortly petiolate, entire, exstipulate, usually ± elliptic,
usually glabrous at maturity, usually with 2–6 lateral nerves curving up
from near the base to become subparallel. Flowers small, numerous, in
spikes. Calyx-lobes (? 3–) 4–5, ± triangular-deltoid, ± erect. Petals 4–5.
Stamens twice as many as the calyx-lobes ; anthers didymous. Ovary
inferior ; styles 4–5 (–8), ± thickened below ; ovules solitary in each cell.
Fruit a ± oblong drupe. Endosperm absent.

The species known from our area also have the following characters in common :

Leaves usually narrowly to broadly elliptic or lanceolate, more rarely
ovate or rotund. Calyx-lobes 4 (–7), glabrous within. Petals 4 (–6), divided
above. Stamens 8 (–12), with ± raised ± convoluted or muricate glands
between their bases. Styles 4 (–5).

A genus widespread in tropical Africa, with a few species in India and the Malay
peninsula and recently discovered in tropical South America. Well characterized by
the leaf shape and venation, and known for the edible fruit.

Trees more than 20 m. tall ; smaller leaves alternating with
       the larger ; leaves papery ; bracts on the rhachis of
       the inflorescence longer than the buds   .   .   .   3. *A. obtusifolia*

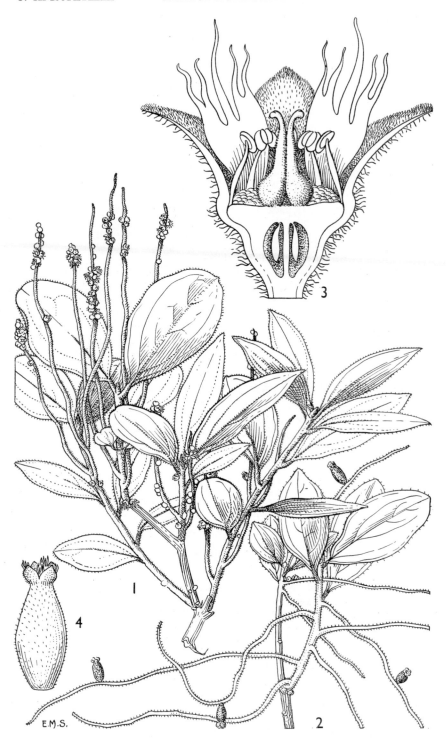

FIG. 5. *ANISOPHYLLEA BOEHMII*—**1,** flowering branch, × ⅔ ;   **2,** fruiting branch, × ⅔ ;   **3,** L.S. flower, × 20 ; **4,** young fruit, × 4. 1, 3 from *Watkins* 156 ; 2, 4 from *Bullock* 3218.

Trees or shrubs less than 12 m. tall ; smaller leaves not
    alternating with larger ; leaves leathery ; bracts on
    the rhachis of the inflorescence shorter than the
    buds :
    Young leaves tomentose, hairs long, stout, thick-walled
        and slightly twisted ; mature leaves tomentose ;
        mature spikes ± densely tomentose .   .   . 1. *A. boehmii*
    Young leaves tomentulose, hairs short, weak, thin-walled
        and moderately twisted ; mature leaves and spikes
        glabrous or nearly so   .   .   .   .   . 2. *A. pomifera*

1. **A. boehmii** *Engl.* in P.O.A. C : 287 (1895) ; V.E. 3, 2 : 675 (1921) ;
T.T.C.L. 470 (1949) ; Duvig. & Dewit in Bull. Inst. Roy. Col. Belge, 21 :
925 (1950). Type : Tanganyika, Tabora District : " Pori on the heights
between Magengwe & Bamanda," *Boehm* 80a (B, holo. †)

    Tree up to 10 m. tall (in our area) or large shrub. Young parts densely
crimson- or yellow-brown tomentose, glabrescent, not glabrate ; hairs
long, stout, ± thick-walled and slightly twisted. Leaf blades on upper parts
of branches narrowly to broadly elliptic, rarely rotund, 6–11 cm. × 1–7 cm.,
leathery, narrowed or subrounded to a cuneate or acute base, narrowed or
rounded to an obtuse apex, longly ± densely tomentose on both surfaces
especially towards base. Spikes 4–9 cm. long, in axils of smaller leaves
(bracts) on lower parts of branches ; bracts obovate, 1–3 cm. long, very
deciduous ; bracts on the rhachis inconspicuous, subterete, 2 mm. long.
Petals 3-lobed, lateral lobes bipartite. Styles ± free, narrowly pyriform,
divergent. Fruits not seen. Fig. 5.

TANGANYIKA. Buha District : Kibondo Road, Malagarasi Pontoon Camp, 27 Aug .
   1950, *Bullock* 3218 ! ; Mwanza District : Ibondo, 31 Mar. 1937, *B. D. Burtt* 6526 !
   Chunya District, Kipembawe, May 1951, *Eggeling* 6087 !
DISTR. T1, 4, 7 ; Northern Rhodesia, Belgian Congo
HAB. *Brachystegia-Isoberlinia* woodland ; 1200–1700 m.

SYN. [*A. ferruginea* Rolfe *ms.* Belgian Congo, *Ringoet s.n.* (K !) & 56 (K !)]
      [*A. rogersii* S. Moore *ms.* Belgian Congo, *Rogers* 26340 (K !)]

VARIATION. The material from Chunya District [including also *Wigg in F.H.* 1183 (K !),
   cited as *A. sp.* in T.T.C.L. 471 (1949)] has broader leaves and stouter, more well-
   developed inflorescences with a more deeply coloured indumentum than that from
   the northern areas. It may represent a distinct variety.

NOTE. The leaves persist throughout the dry season, but fall when the new shoots
   develop.
     A. fruticulosa Engl. & Gilg, from Angola, is very similar in its morphology but
   differs in being a low shrub.

2. **A. pomifera** *Engl. & Brehm.* in E.J. 54 : 376 (1917) ; T.T.C.L. 471
(1949) ; Duvig. & Dewit in Bull. Inst. Roy. Col. Belge, 21 : (1950). Type :
Tanganyika, Rungwe District, Kyimbila, Mulinda Forest, *Stolz* 1453 & 1802
(both K, iso-syn. !)

    Small tree or shrub. Young parts brown-yellow-tomentulose, glabrescent
or glabrate ; hairs short, thin, thin-walled and moderately twisted. Leaf-
blades (on upper parts of branches) ± elliptic, 4·5–6·5 × 2·5–3·2 cm.,
leathery, cuneate below, narrowed to an obtuse apex, shortly tomentose,
especially beneath, very glabrescent, glabrate. Spikes 5–12 (–20) cm. long,
in axils of smaller leaves (bracts) on lower parts of branches ; bracts obovate
1–3 cm. long, deciduous ; bracts on the rhachis inconspicuous, subterete,
± 2 mm. long. Petals 5-lobed. Styles ± free, pyriform, ± reflexed above.
Fruits c. 4 × 2·5 cm. when dry, edible, flavoured like a nectarine.

TANGANYIKA. Rungwe District : Mulinda Forest, 22 July 1912 (fl.), *Stolz* 1453 ! & 3 Jan. 1913 (fr.), *Stolz* 1802 !

DISTR. **T7** ; Northern Rhodesia and Nyasaland

HAB. *Brachystegia-Isoberlinia* woodland ; 1200–1600 m.

3. **A. obtusifolia** *Engl. & Brehm.* in E.J. 54 : 372 (1917) ; V.E. 3, 2 : 675 (1921) ; T.T.C.L. 470 (1949). Type : Tanganyika, Lushoto District, E. Usambara Mts., Derema, *Scheffler* 228 (K, iso-syn. !) & Amani, *Zimmermann* 281 & 2938 (both EA, iso-syn. !)

Very tall tree, up to 45 m. in height. Young parts ± longly hirsute, hairs shining -white or -(brownish-)yellow. Larger leaf-blades narrowly elliptic to elliptic, 2–7·5 × 0·5–3·5 cm., papery, ± rounded or cuneate to a small deltoid base, narrowed to a subacuminate acute apex, glabrous or nearly so at maturity ; smaller leaves alternating with larger, 0·5–1·5 cm. long and relatively broader, otherwise similar. Spikes up to 4 cm. long ; bracts on the rhachis evident, narrowly lanceolate to almost lanceolate, 2–6 mm. long. Petals 3–5-lobed ; lobes with longly attenuate coiled apices. Styles partially joined, inflated below the middle ; narrower apical part very short and ± divergent. Fruits not seen.

TANGANYIKA. Lushoto District : Amani, 24 Jan. 1933, *Greenway* 3336 ! & Sangarawe, 1 Jan. 1937, *Greenway* 4813 !

DISTR. **T3** ; known only from the eastern Usambara Mts.

HAB. Locally common in evergreen rain-forest ; *c.* 900 m.

NOTE. This is a very distinct, almost anomalous species. The fruits have not been observed ; unless their nature falls within the degree of variation shown by the drupes of the other species of *Anisophyllea*, this species may be segregated from our other two species at the generic or subgeneric level. The variety of *A. laurina* mentioned in V.E. 1, 1 : 296 (1910) almost certainly refers to this species, as suggested by Mr. J. P. M. Brenan, T.T.C.L. 470 (1949).

INDEX TO RHIZOPHORACEAE